This book belongs to:

....

....

it's....o'clock

it's five to...   it's five past...

it's ten to..   it's ten past..

it's quarter to..   it's quarter past

it's twenty to..   it's twenty past..

it's twenty five to...   it's twenty five past..

it's half past...

this hand shows minutes   this hand shows hours

example

it's 1 o'clock   it's half past 1

it's five past 1   it's twenty five to 2

it's ten past 1   it's twenty to 2

it's quarter past 1   it's quarter to 2

it's twenty past 1   it's ten to 2

it's twenty five past 1   it's five to. 2

When the big hand is on the twelve, we say it's o'clock
exercice1 : write the time below

☐ o'clock ☐ o'clock ☐ o'clock ☐ o'clock

☐ o'clock ☐ o'clock ☐ o'clock ☐ o'clock

☐ o'clock ☐ o'clock ☐ o'clock ☐ o'clock

When the big hand is on the three, we say it's quarter past.
exercice2 : write the time below

quarter past ☐   quarter past ☐   quarter past ☐   quarter past ☐

quarter past ☐   quarter past ☐   quarter past ☐   quarter past ☐

quarter past ☐   quarter past ☐   quarter past ☐   quarter past ☐

When the big hand is on the six, we say it's half past.
exercice3 : write the time below

half past ☐   half past ☐   half past ☐   half past ☐

half past ☐   half past ☐   half past ☐   half past ☐

half past ☐   half past ☐   half past ☐   half past ☐

When the big hand is on the nine, we say it's quarter to

exercice4 : write the time below

quarter to ☐   quarter to ☐   quarter to ☐   quarter to ☐

quarter to ☐   quarter to ☐   quarter to ☐   quarter to ☐

quarter to ☐   quarter to ☐   quarter to ☐   quarter to ☐

exercice 5 : match the time

it's twenty five to 9 • • 1:30

it's half past 1 • • 1:55

it's ten past 5 • • 7:05

it's five to. 2 • • 5:10

it's twenty past 3 • • 3:45

it's twenty five past 8 • • 3:20

it's five past 7 • • 6:00

it's 6 o'clock • • 9:15

it's twenty to 5 • • 8:35

it's quarter to 4 • • 8:25

it's ten to 3 • • 4:40

it's quarter past 9 • • 2:50

exercice 6 : match

 •   it's ten o'clock   • 12:35

 •   it's twenty five to one   • 10:00

 •   it's half past nine   • 10:40

 •   it's twenty to eleven   • 9:30

exercice 7 : look and write the time

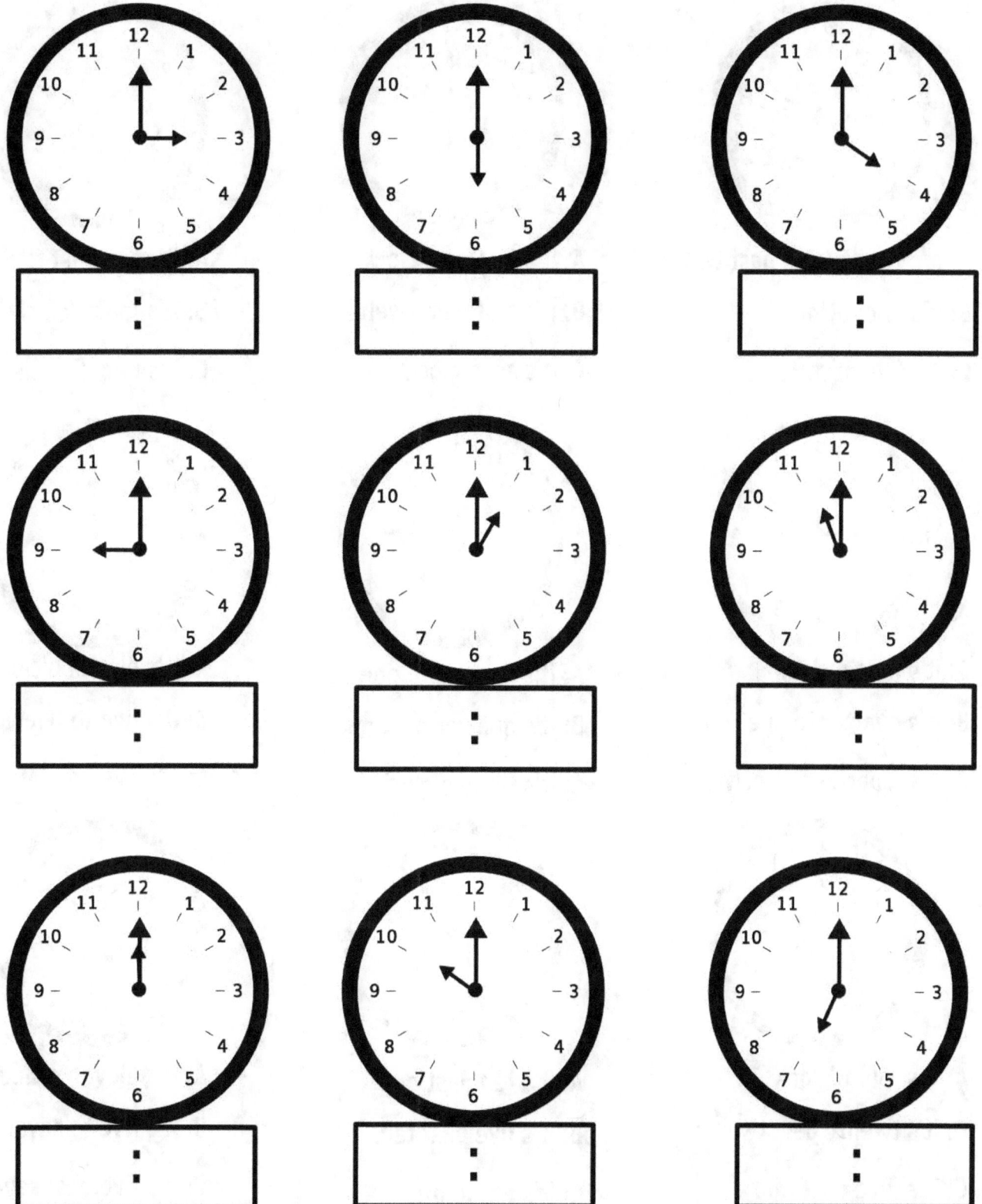

exercice 8 : circle the right answer

A: it's twenty five past two

B: it's two o'clock

C: it's five past two

A: it's twelve o'clock

B: it's half past twelve

C: it's six o'clock

A: it's ten past four

B: it's four o'clock

C: it's two o'clock

A: it's nine o'clock

B: it's quarter to one

C: it's quarter to twelve

A: it's quarter to one

B: it's quarter past one

C: it's three o'clock

A: it's eleven o'clock

B: it's five to twelve

C: it's five past eleven

A: it's four o'clock

B: it's twenty past twelve

C: it's five past four

A: it's ten past one

B: it's five past ten

C: it's ten to one

A: it's eleven o'clock

B: it's five to seven

C: it's five past seven

exercice 9 : circle the right answer

A: 5:11
B: 4:55
C: 5:55

A: 6:00
B: 10:06
C: 6:10

A: 6:25
B: 5:30
C: 6:05

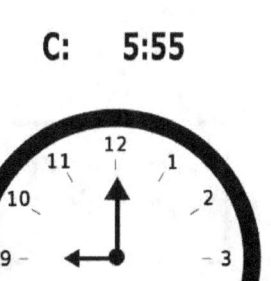

A: 12:09
B: 12:45
C: 9:00

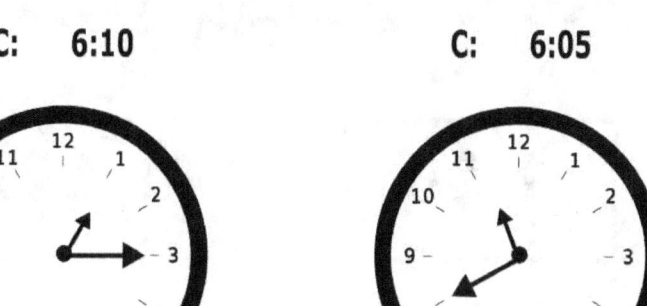

A: 1:15
B: 1:03
C: 3:05

A: 10:40
B: 11:40
C: 12:40

A: 4:05
B: 1:20
C: 1:40

A: 2:50
B: 10:15
C: 3:10

A: 8:11
B: 7:55
C: 8:55

exercice 10 : color the right answer

it's half past one

| 1:30 | 1:15 | 3:10 |

it's one o'clock

| 1:00 | 12:00 | 1:10 |

it's five past ten

| 10:05 | 5:10 | 10:50 |

it's eleven o'clock

| 11:00 | 11:10 | 11:30 |

exercice 11: color the right answer

2:30

| it's half past two | it's quarter to two | it's half past one |

10:10

| it's ten to ten | it's ten past ten | it's ten to eleven |

5:15

| it's five past five | it's quarter to five | it's quarter past five |

3:45

| it's quarter to four | it's quarter to three | it's quarter past three |

exercice 12: draw the clock hands

11:15	12:30	1:20

6:45	**7:30**	**3:05**

4:15	**3:30**	**10:10**

exercice 13: write the time in numbers

it's ten past eleven

....

it's quarter to four

....

it's three o'clock

....

it's quarter past nine

....

it's half past twelve

....

exercice 14: write the time

12:30

....

4:50

....

8:15

....

7:45

....

10:00

....

exercice 15 : draw the taller hand

11:05

9:00

1:20

4:20

6:30

6:45

12:15

8:25

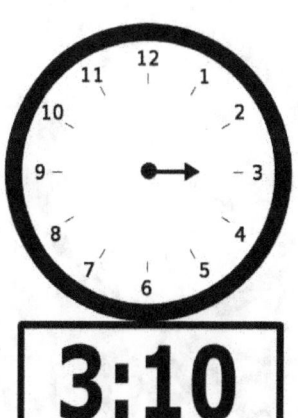
3:10

exercice 16: draw the shorter hand

12:25

9:35

12:50

2:15

3:30

6:00

5:40

1:45

4:25

exercice 17: color or circle the quarter past

12:15	3:40	2:30	6:10	1:15
3:30	5:15	1:50	7:15	5:15
8:45	7:20	2:30	5:30	7:20
7:40	1:05	1:15	9:40	1:30
5:10	8:15	7:50	3:15	8:45
10:15	6:15	1:30	7:50	11:50
6:45	8:45	3:15	8:15	12:15
8:15	10:20	5:10	9:05	3:30
12:05	1:30	8:15	10:30	4:50
2:15	2:15	11:30	3:20	8:15

exercice 18: color or circle the quarter to

1:30	3:40	2:30	6:10	1:15
8:45	5:15	1:50	7:15	5:15
11:50	7:20	2:30	5:30	9:05
7:40	1:05	1:15	9:40	10:30
7:20	3:30	7:50	3:15	3:20
12:15	4:50	1:30	7:50	8:45
3:30	8:15	3:15	8:15	12:15
8:15	10:20	5:10	8:15	5:10
12:05	1:30	8:15	6:15	10:15
2:15	2:15	11:30	8:45	6:45

exercice 19: color or circle the half past

1:30	3:40	2:30	6:10	1:15
8:30	5:15	1:50	7:15	5:15
11:50	7:20	2:30	5:30	9:05
7:40	1:05	1:15	9:40	10:30
7:20	3:30	7:50	3:15	3:20
12:15	4:50	1:30	7:50	8:45
3:30	8:15	3:15	8:15	12:15
8:15	10:20	5:10	8:30	5:10
12:05	1:30	8:15	6:15	10:15
2:15	2:15	11:30	8:45	6:45

exercice 20: color or circle the " ...o'clock "

1:00	3:40	2:30	6:10	1:00
8:30	5:00	1:50	7:15	5:15
11:50	7:20	2:00	5:30	9:05
7:40	1:05	1:00	9:40	10:30
7:20	3:30	7:50	3:15	3:20
12:15	4:50	1:30	7:50	8:45
3:30	8:15	3:15	8:15	12:15
8:00	10:20	5:10	8:30	5:10
12:05	1:30	8:15	6:15	10:15
2:15	2:15	11:30	8:00	6:45

exercice 21: find the followings

_it's half past one
_it's quarter to nine
_it's ten past six
_it's quarter past three
_it's ten to two

1:00	3:40	2:30	6:10	1:00
8:30	5:00	1:50	7:15	5:15
11:50	7:20	2:00	5:30	9:05
7:40	1:05	1:00	9:40	10:30
7:20	3:30	7:50	3:15	3:20
12:15	4:50	1:30	7:50	8:45
3:30	8:15	3:15	8:15	12:15
8:00	10:20	5:10	8:30	5:10
12:05	1:30	8:15	6:15	10:15
2:15	2:15	11:30	8:00	6:45

exercice 22: find the followings

4:30
7:00
9:30
6:45
1:15
7:55

it's eleven o'clock	it's half past four
it's half past nine	it's quarter past three
it's quarter to seven	it's ten to two
it's quarter past one	ten past two
it's twenty five to one	it's seven o'clock
it's quarter past ten	it's five to eight

exercice 23: look at the clock and write the right answer both in words and numbers

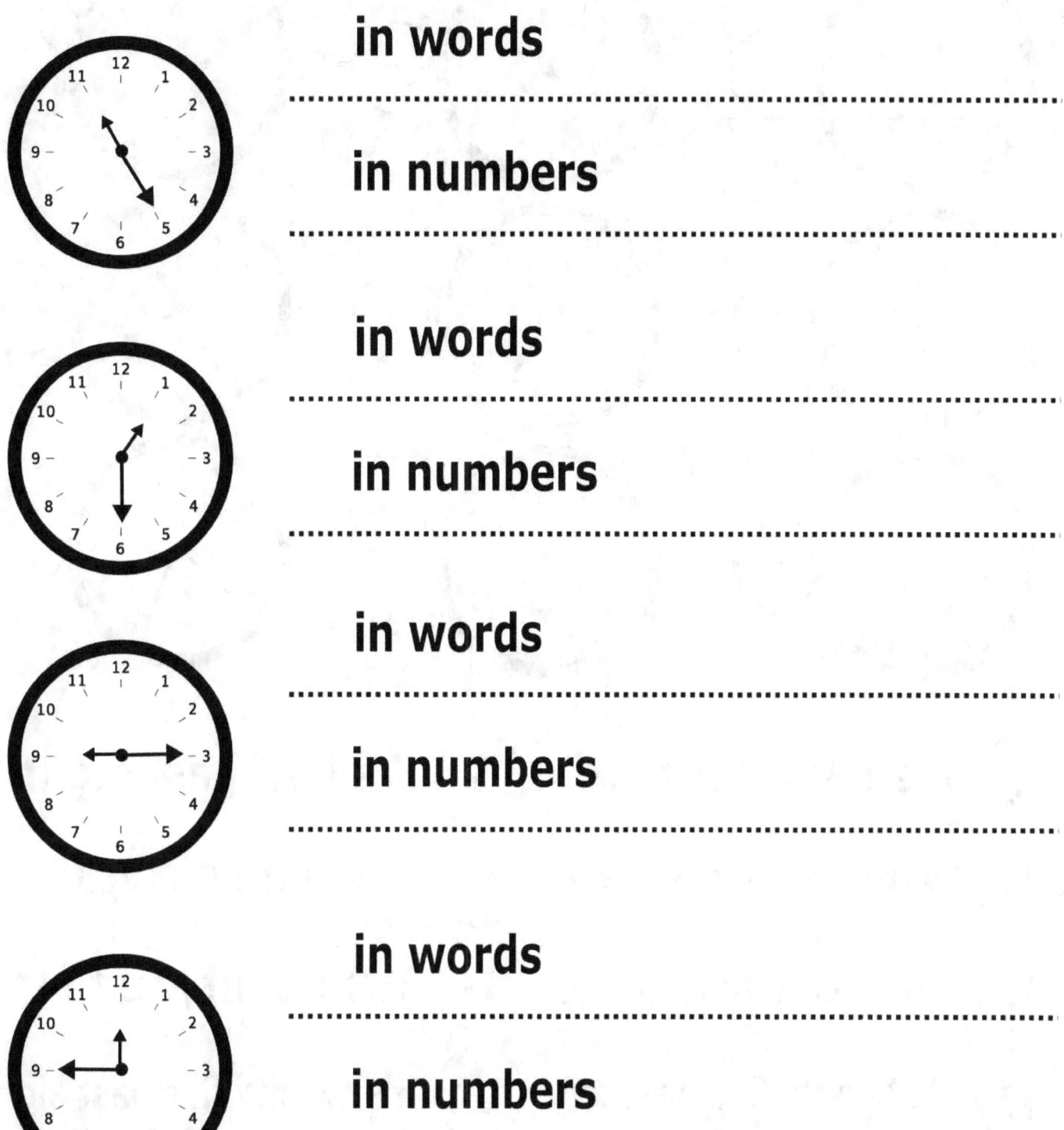

exercice 24: what time is it ?

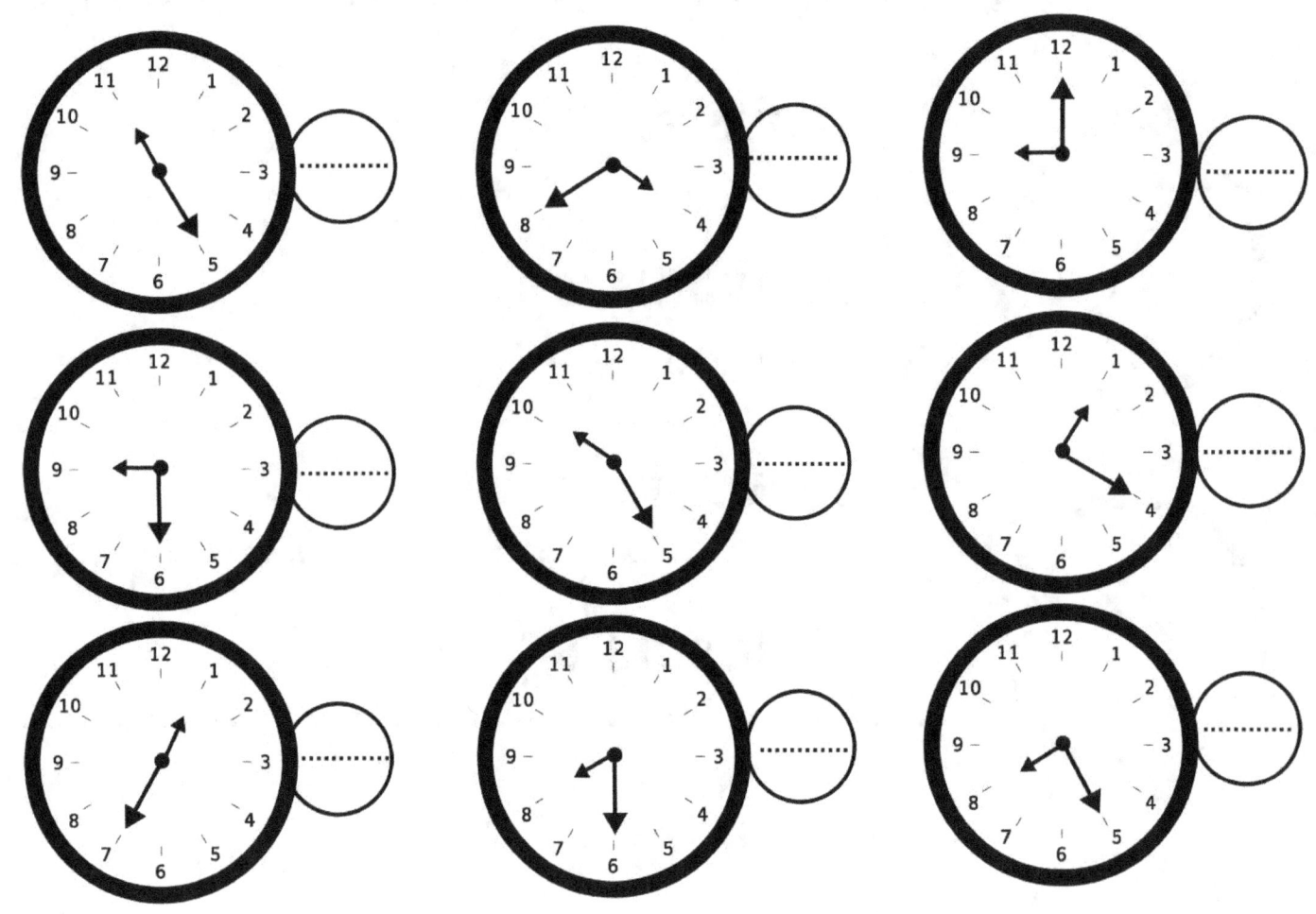

- **A)** it's half past nine
- **B)** it's twenty five past eleven
- **C)** it's twenty five to one
- **D)** it's twenty five past ten
- **E)** it's twenty past one
- **F)** it's half past eight
- **G)** it's nine o'clock
- **H)** it's twenty to four
- **I)** it's twenty five past eight

exercice 25: tick the right answer

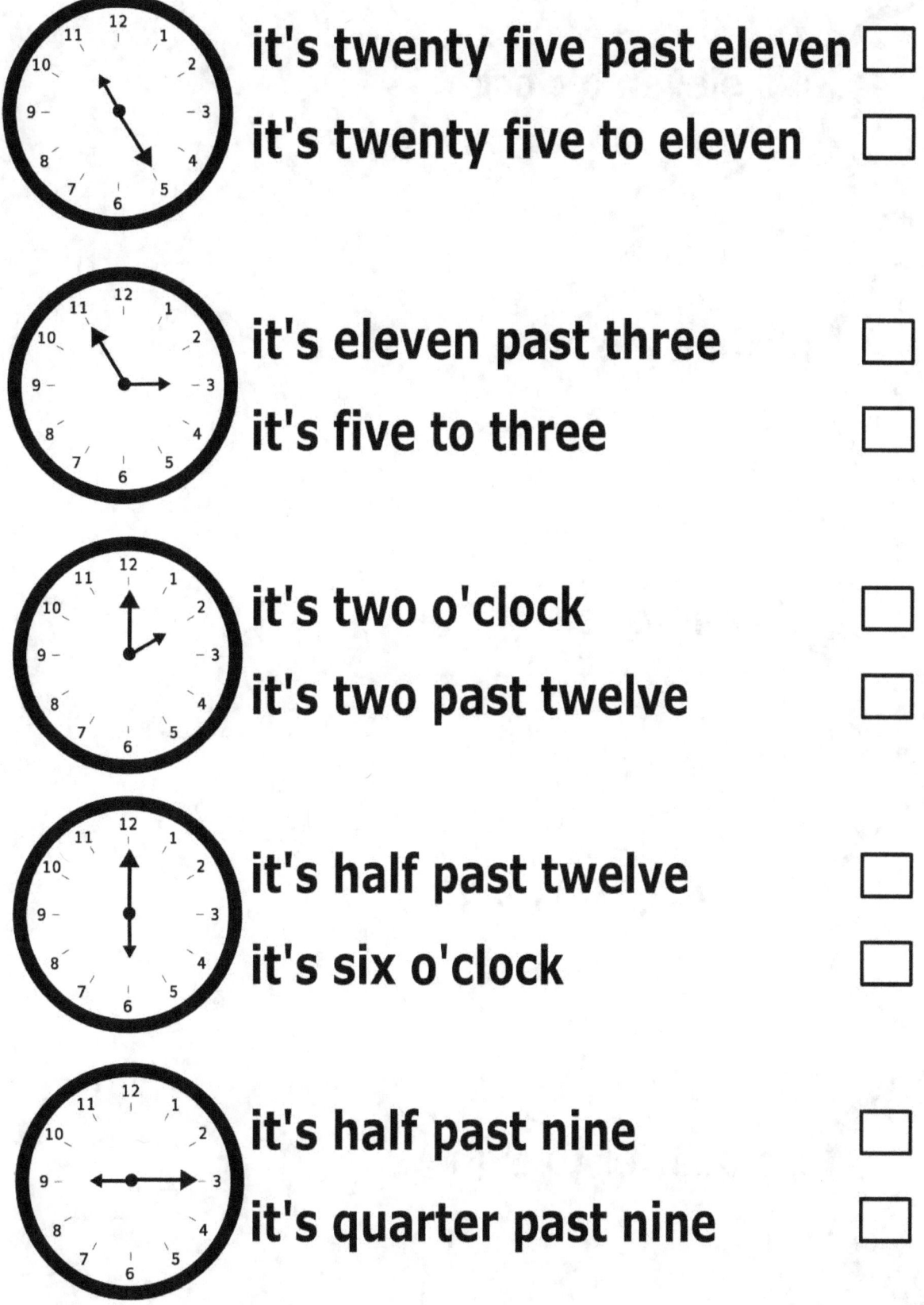

exercice 26: write the right number and draw the hands

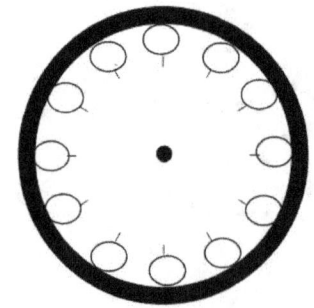 it's eleven o'clock

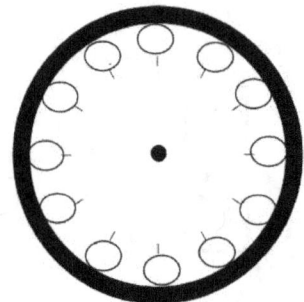 it's half past one

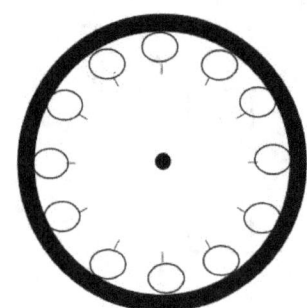 it's ten to ten

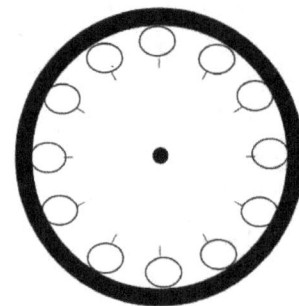 it's five past twelve

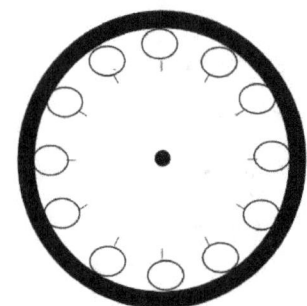 it's quarter to eight

exercice 27: look at the time and Fill in the blanks

7:15 → it's ....seven

9:00 → it's nine ....

4:45 → it's quarter ....

2:15 → it's quarter ....

8:10 → it's .... eight

7:50 → it's .... eight

3:30 → it's ....

5:55 → it's five ....

9:45 → it's..............................ten

1:10 → it's ten..................

exercice 28: is it " quarter to " or quarter past"

8:45 →..

5:15 →..

3:15 →..

12:15 →..

4:45 →..

6:15 →..

3:45 →..

2:15 →..

4:15 →..

8:45 →..

exercice 29: true or false

10:20 → it's twenty to ten	
3:15 → it's quarter to three	
4:00 → it's half past four	
11:00 → it's eleven o'clock	
10:10 → it's ten past ten	
6:15 → it's quarter to six	
8:00 → it's half past eight	
11:35 → it's twenty five to twelve	
2:40 → it's twenty to three	
1:30 → it's half past one	

exercice 30: observe and fill

it's half past............. | 9:...... |

it's ....eleven | ......:15 |

it's quarter past........ | 12:...... |

it's quarter to .... | 1:...... |

it's twenty to ..... | 11:..... |

it's seven........ | ......:00 |